LAYERS OF LIQUIDS

Written by
Read With You Science Academy

Read With You

The characters and events portrayed in this book are fictitious. Any similarity to real persons, living or dead, is coincidental and not intended by the author.

Published by Read With You Publishing
Designed by Read With You Science Academy
Read With You and associated logos are trademarks and/or registered trademarks of Read With You L.L.C.

ISBN-13: 978-1-64616-113-3
ISBN-10: 1-64616-113-0

Second Edition April 2021
Printed in the United States of America.

Welcome to the

READ WITH YOU SCIENCE ACADEMY!

Today we will do an experiment called

LAYERS OF LIQUIDS

Before we begin, let's review some important words:

volume:
The amount of space that is filled by something.

density:
The amount of mass in a known volume of an object or substance.

To mix:
To combine two or more things to make one thing that is the same throughout.

SAFETY FIRST

Remember:
Be a careful scientist! Clean up any water that spills right away! Remember to use a bowl or a tray to collect spills.

QUESTION:

What I want to know.

Will all liquids mix together?

- [ ] Yes, all liquids will mix together.
- [ ] No, all liquids will not mix together.

HYPOTHESIS:

What I think will happen and why.

All liquids **will** mix together evenly because...

All liquids **will not** mix together evenly because...

All liquids **will** mix together evenly because:	All liquids **will not** mix together evenly because:
Each liquid has the same density as the other.	Each liquid has a different density than the other.
All liquids are the same.	Less dense liquids float.

What do you think?
Circle your hypothesis!

EXPERIMENT:

How we will test our hypothesis.

THINGS WE NEED

- A tall, clear drinking glass
- A large spoon
- Water
- 1 cup of vegetable oil
- Syrup (Corn syrup if available)
- Food coloring
- Paper towels

water

vegetable oil

glass

syrup

spoon

food coloring

PROCEDURE:

The steps to complete the experiment.

Step 1A

- Fill the glass halfway with plain water.
- Remember to put the glass over a bowl or in a pan in case your **liquid** spills!

Step 2A

- Add just one drop of food coloring.
- Use a paper towel to clean up any spills.

Step 3A

- Slowly pour the syrup into the water and **observe** what happens. Does the syrup sink or float?

Step 4A

- Hold the spoon just above the surface of the water. It should be touching the side of the glass. Then, slowly pour the oil onto the spoon.
- Does the oil sink or float?
- Remember: Use a towel or paper towel to clean any spills!

Write down what you see

In your science journal,

write what you **observed**!

- What was your **hypothesis**? (Look on page 5 of this book!)
- Was your hypothesis correct?
- Did the oil sink or float?
- Did the syrup **mix** with the water?

Let's do it again!

What will happen if we add the food coloring last? Let's do an experiment to find out. You will need the same materials as the first experiment.

QUESTION:

What I want to know.

What will happen if the food coloring is added last?

HYPOTHESIS:

What I think will happen and why.

I **think** adding the food color last will change the experiment because…

I **don't think** adding the food color last will cause any change because...

What do you think?

Write your new hypothesis in your science journal.

PROCEDURE:
The steps to complete the experiment.

Step 1B

- Fill the glass halfway with plain water.
- Remember to put the glass over a bowl or in a pan in case your liquid spills!

Step 2B

- Slowly pour the syrup into the water and observe what happens.

Step 3B

- Hold the spoon just above the surface of the water. It should be touching the side of the glass. Then, slowly pour the oil onto the spoon.
- Be a careful scientist and keep your work space dry!

Step 4B

- Add two drops of food coloring.
- Observe what happens.

Write down what you see

In your science journal,

write your results!

- What was your hypothesis? (Look on page 11 of this book!)
- Was your hypothesis correct?
- What happened when the food coloring was added last?

Draw a picture of what you saw when the oil and syrup were added.

Why did this work?

- This experiment worked because each of the liquids we used has a different **density**.
- The corn syrup was the densest liquid, so it sank to the bottom of the glass.
- Water has a greater density than vegetable oil, but it is less dense than corn syrup. That's why it was the middle layer.
- Finally, the vegetable oil had the least amount of density of the three liquids we used, so it formed the top layer.

SCIENCE

- We used this experiment to see how liquids of differing densities interact with one another.
- Density is a measurement that compares the amount of matter an object has to its **volume**. If **mass** is a measure of how much "stuff" there is in an object or liquid, we can think of density as a measure of how tightly that "stuff" is packed together. For example, the corn syrup is filled with lots of sugar, so it has a lot of "stuff." The "stuff" is very close together, which means the corn syrup is very dense. That's why it sunk to the bottom of the glass.
- The "stuff" in water is closer together than the "stuff" in vegetable oil, so it is denser than the vegetable oil. Water is still less dense than the corn syrup, though. That's why it formed the middle layer. Of the three liquids, the "stuff" in vegetable oil is packed together less than in the water or corn syrup, so it is the least dense liquid and forms the top layer.

GLOSSARY

Density: A measurement that compares the amount of matter an object has to its volume.

Liquid: A substance that is able to flow freely.

Mass: The quantity of matter in something.

To mix: To combine two or more things to make one thing that is the same throughout.

Hypothesis: An educated guess; what we think will happen and why.

To observe: To watch something carefully.

Procedure: The steps to complete the experiment.

Question: What we want to know.

Volume: The amount of space that is filled by something.

NEW IDEAS

How could we change this experiment?

- What would happen if we reversed the order of the liquids?
- What would happen if we added liquid soap?

Can you think of any ideas to improve the experiment?

- What would happen if we used a water bottle, closed the top, and shook it?

JOURNAL

Made in United States
Orlando, FL
07 November 2026

90337521R00017